Table of Contents

Executive Summary..1
Introduction..3
Major Accomplishments for 2003-2008.......................................5
Plan for 2009-2013..9
Appendix..18

Executive Summary

In 1998 the National Plant Genome Initiative (NPGI) was established as a coordinated national research program by the Interagency Working Group (IWG) on Plant Genomes with representatives from the U.S. Department of Agriculture (USDA), U.S. Department of Energy (DOE), National Institutes of Health (NIH), National Science Foundation (NSF), Office of Science and Technology Policy (OSTP), and Office of Management and Budget (OMB). The current IWG also includes the U.S. Agency for International Development (USAID). Since 1998, the field of plant genomics has been revolutionized by the freely accessible tools and resources developed through the NPGI. The NPGI has changed the way research is conducted in plant biology and beyond; it has attracted a new generation of scientists to plant research; and it has contributed new knowledge and ideas to science. Through the development of plant genomic and other "-omics" resources, the NPGI has built a foundation on which the scientific community is advancing research, not just in plant genomics but across diverse disciplines spanning the biological sciences.

This report describes the NPGI plan for the next five years (2009 – 2013). The IWG based this plan on a wide range of inputs from the broader community which are listed in the Appendix.

NPGI Goal Statement

The goal of the NPGI is to develop a basic knowledge of the structures and functions of plant genomes and translate this knowledge to a comprehensive understanding of all aspects of economically important plants and plant processes of potential economic value. By bridging basic research and plant performance in the field, the NPGI will accelerate basic discovery and innovation in economically important plants and enable enhanced management of agriculture, natural resources, and the environment to meet societal needs.

Guiding Principles for NPGI, 2009-2013

The NPGI will follow the same guiding principles used for the first ten years:

- The Initiative should be a long-term project governed by a periodically updated plan based on scientific progress and with stakeholder input as a basis for setting goals and priorities

- All research resources, including data, software, germplasm, and other biological materials and research tools should be openly accessible to all in a timely manner

- The Federal portion of the Initiative should be coordinated by an interagency working group

- All awards should be made based on scientific merit and rigorous, competitive peer review

- Partnerships with the private sector and the international community are vital for success

New Objectives for 2009-2013:

Objectives for the next five years of the NPGI will build on recent scientific and technical discoveries to ensure continued advancement in plant genomics specifically and plant sciences in general. The major objectives are to:

- Expand genomic resources for every major plant of economic importance

- Advance plant systems biology

- Translate basic discovery to the field

- Develop coordinated solutions to data access, data analysis and data synthesis

- Enhance education, training and outreach

- Broaden societal impacts

To capitalize fully on the investments made to date, to enable further advances in plant genome research and to address the impacts of global climate change, the agencies participating in the NPGI will continue efforts in the context of overall Administration and agency priorities and available resources. The funds will continue to be expended on a competitive basis using rigorous peer review. As in the first ten years of the NPGI, the IWG will monitor progress and report major accomplishments annually.

I. Introduction

The Interagency Working Group (IWG) on Plant Genomes was appointed in May 1997 by the Executive Office of the President in response to a request from the Senate VA, HUD, and Independent Agencies Appropriations Subcommittee. The IWG consisted of representatives from the U.S. Department of Agriculture (USDA), National Science Foundation (NSF), National Institutes of Health (NIH), U.S. Department of Energy (DOE), Office of Science and Technology Policy (OSTP), and Office of Management and Budget (OMB). The IWG was charged with identifying science-based priorities for a national plant genome initiative and to plan and coordinate plant genome research activities for the Nation. The IWG, which now includes a representative from the U.S. Agency for International Development (USAID), continues to play a key role in the coordination of plant genome research across the federal government as well as with industrial and international partners.

The IWG published the first National Plant Genome Initiative (NPGI) five-year plan in January 1998[1] and the second five-year plan in January 2003[2]. The second five year plan (National Plant Genome Initiative: 2003-2008) recommended the following plant genome research objectives:

- Contribute to the international effort to finish the rice genome sequence
- Complete sequencing of the gene-rich regions of the maize genome
- Complete the Arabidopsis functional genomics project as a Rosetta Stone for all plants
- Establish repositories for plant genome research resources
- Apply genomics tools to understanding traits of biological or economic importance, including wood formation, fruit development, and nitrogen metabolism
- Expand genomics approaches to biodiversity, ecology, and ecosystems studies as well as development of renewable resources
- Develop informatics tools to access and use plant genome databases

The accomplishments over the past five years have more than surpassed these objectives. This acceleration of advances was made possible by investments in research resources over the ten years of the NPGI, which in turn, enabled researchers to address major unanswered questions in plant biology as well as translate basic research into advances in the field. The development of next-generation sequencing machines through NIH support and increased availability of computational capability to individual laboratories

1   http://www.ostp.gov/galleries/NSTC%20Reports/PlantGenomeInitiative1998.pdf
2   http://www.ostp.gov/pdf/npgi2003_2008.pdf

are transforming the way that plant genomics research is conducted. This has broadened the participation in plant genomics research across a wider range of institutions and posed new challenges for data analysis and data management.

The success of the NPGI over its ten-year history has been due in part to the strict adherence to the guiding principles set out in the first five-year plan:

- The Initiative should be a long-term project governed by a periodically updated plan based on scientific progress and with stakeholder input as a basis for setting goals and priorities
- All research resources, including data, software, germplasm, and other biological materials and research tools should be openly accessible to all in a timely manner
- The Federal portion of the Initiative should be coordinated by an interagency working group
- All awards should be made based on scientific merit and rigorous, competitive peer review
- Partnerships with the private sector and the international community are vital for success

In this report, the IWG documents significant NPGI achievements since 2003 and articulates a new plan for the next five years (2009-2013). The IWG solicited and received input from many sources while developing this plan, including a report from the National Academy of Sciences and a variety of stakeholder workshops involving scientists, growers, producers, and the public. These inputs were used to identify the scientific opportunities and challenges for the NPGI in the next five years (see Appendix). The new plan will be used by each of the participating agencies to establish priorities and initiate new activities consistent with the mission of each agency and with the overall goals of the NPGI.

Photo Removed Due to Copyright Restrictions

II. Major Accomplishments for 2003-2008

The accomplishments made possible through the activities of the NPGI over the past 10 years have been spectacular, having impacts across the public sector, in industry, and internationally. The funding provided by NPGI is leading to increased participation in plant genome research and generating high impact publications. In a recent survey conducted by the National Research Council[3], 165 investigators supported through the NPGI cited 1,478 peer-reviewed publications included in the 2006 ISI Journal Citation Impacts, 21% of which were in the most highly cited journals. In short, under the NPGI U.S. researchers developed a greater understanding of the basic biology of plants. This knowledge has yielded societal benefits and allowed the U.S. to remain in the forefront of innovation in a plant-based economy. Activities of the NPGI provided the foundation for yield increases in corn, development of plant-based biofuels and new biomaterials, and contributed to improved food security.

In the previous Five-Year Plan, the IWG set a number of ambitious scientific objectives, which have been met or exceeded. Highlights of the scientific and societal impacts include:

Discovery

- A detailed understanding of the sequence and structure of the Arabidopsis and rice genomes, the encoded genes, and many of their functions.

  o The completed **rice genome sequence**, released in December 2004, has continued to be refined and updated, with annotation of over half the 50,000 predicted genes and the surprising discovery of gene activity buried within the spans of repetitive, non-coding DNA that typically comprises a native centromere[4].

  o Investments in the internationally-coordinated **Arabidopsis functional genomics** efforts identified a wealth of new gene and gene functions. Outcomes from this work are being translated into a deeper understanding of gene networks in crop plants as well as the development of new varieties.

- A deeper understanding of the **impacts of domestication** on the genome organization and gene content of major U.S. crops such as maize, cotton and wheat. For example, domestication of maize involved changes in expression of a small number of genes that impacted shoot branching and flowering.

---

3   The National Academy of Sciences report entitled "Achievements of the National Plant Genome Initiative and New Horizons in Plant Biology", January 29, 2008. (http://www.nap.edu/catalog.php?record_id=12054).
4   A centromere is a highly condensed region of a chromosome visible as a constriction. It serves as the attachment point for the cellular machinery that separates sister chromosomes during cell division.

Tool Development

- **Completion of the draft sequence of the maize genome** in February 2008 is already accelerating efforts in maize genetics, genomics and breeding to meet society's growing demands for food, feed, and biofuels.

- Development of additional **physical maps** for wheat, soybean, peach, apple, and common bean are fundamental to the development of DNA markers, Quantitative Trait Locus (QTL) mapping, positional cloning of genes, as well as functional and comparative genomics.

- **Completion of genome sequences for a diverse set of plant species** accelerated dramatically between 2003-2008, with the release of draft annotated genomes of poplar, sorghum, soybean, *Medicago*, and two mosses (*Physcomitrella* and *Selaginella*) joining the *Arabidopsis* and rice genomes. Significant technology development and instrumentation were requisite to overcome each species' unique challenges with respect to genome structure, including the extent and location of duplications, rearrangements, expansions, and gene density.

- Development of **computational tools** to elucidate the relationships between physical traits and genes in plants in a high-throughput and cost-effective way.

- **Databases** for comparative cereal genomics (Gramene), comparative legume genomics (Legume Information System), the Triticeae genomes such as wheat, barley and rye (GrainGenes), and maize genomics (MaizeGDB) now provide a one-stop shop for genomic data and resources for major crop plants. In addition, specialized databases now provide a portal for plant expression data (PlexDB) and comparative plant genomics (PlantGDB)[5].

Translation

- Advances in **Marker-Assisted Selection** tools, high-throughput genome sequencing and genotyping as well as quantitative and population genetics have led to enhanced uses of association mapping for improved plant production and protection. For example, a simple sequence repeat (SSR) genetic marker was identified that is 99.2% accurate in identifying soybean breeding lines that carry resistance to Asian soybean rust at the Rpp1 resistance locus. Asian soybean rust caused by *Phakopsora pachyrhizi* was first found in the continental United States in 2004 and is a significant threat to production of soybean in the U.S. This SSR marker will be useful for integrating Rpp1 resistance into modern cultivars.

- **Coordinated Agricultural Projects (CAPs)** in rice, wheat, barley, and conifers

---

5 Links to individual databases are provided in the Appendix

are linking laboratories and field research to provide breeders with new tools to enhance and accelerate traditional crop and tree improvement activities. For example, the Wheat CAP has empowered 20 U.S. public wheat breeders to incorporate modern marker assisted selection techniques into their programs. Marker-assisted selection has been conducted in partnership with the regional small grains genotyping laboratories. These activities have already resulted in the release of 50 improved germplasm and varieties, including new germplasm with promising resistance to cereal rusts and improved end-product quality. Incorporation of modern selection technologies is essential to maintain the international competitiveness of U.S. wheat.

Innovation

New research partnerships have led to innovative approaches to addressing urgent needs, including development of plant-based bioenergy applications and feedstocks and new ways of analyzing data.

- Establishment of three **DOE Bioenergy Research Centers**, which are pursuing basic research underlying a range of high-risk, high-return biological solutions for bioenergy applications. New scientific knowledge generated by these centers will help lay the foundation for biobased products, methods, and tools that the emerging biofuel industry can use. Each center is a multidisciplinary partnership with expertise spanning the physical and biological sciences, including genomics, microbial and plant biology, analytical chemistry, bioinformatics, and engineering.

- Initiation of a **Plant Feedstock Genomics for Bioenergy** Program, which has supported a diverse portfolio of genomic research projects in poplar, sorghum, maize (corn), switchgrass, alfalfa, wheat, perennial grasses, and model bioenergy crops. The knowledge obtained from these projects will provide the scientific foundation to facilitate and accelerate the use of woody plant tissue for bioenergy and biofuels.

- The new **iPlant Collaborative** is a distributed, cyberinfrastructure-centered project designed to bring together and enable the community to tackle grand challenge questions across all of plant biology, including ecological, evolutionary, organismal, molecular, cellular, and developmental areas. An important component of the iPlant Collaborative's activities is training in computational thinking at all levels that is closely integrated with the research activities.

Broader Impacts

- There has been a substantial investment in **undergraduate, graduate, and**

postdoctoral training. One hundred and sixty five NPGI-supported researchers that responded to a recent survey[6] reported training of 1496 undergraduates, 547 graduate students and 717 postdoctoral researchers, many of whom are now in academia, industry, and government.

- **A diverse set of training and outreach activities**, including "Classroom Activities in Plant Biotechnology", "Workshops For 5th -12th Grade Teachers of Science and Agriculture", workshops on "DNA Markers, Mapping, and Beyond", and the "Secrets of Plant Genomes: Revealed!"[7] is enabling research advances made through the NPGI to have an impact on K-12, community college, and university education as well as for end-users such as breeders and growers.

- **Participation in internationally coordinated collaborative activities** enabled by the NPGI, including the projects focused on the Solanaceae, wheat, poplar and rice.

- The **Developing Country Collaborations in Plant Genome Research** activities, which began in 2004, allow the advances made through the NPGI to be expanded to developing country crops and agricultural problems. Achievements include development of breeder-friendly maps for legume crops of local importance in India and Nigeria, including chickpea, pigeonpea, and cowpea as well as development of a rice breeding course offered annually at the International Rice Research Institute in the Philippines.

---

*Cultivation of submergence tolerant rice in Asia is expected to provide protection against damaging floods and increase world food security for resource limited farmers. An international team of researchers cloned the **submergence tolerance gene from rice** and developed a new rice variety that is now producing dramatic gains in yield in farmers' fields in Bangladesh, an area of the world where families live on less than $1/day.*

In summary, the NPGI has advanced basic science while attracting new researchers to plant research and training a new generation of scientists. It has also enabled more effective translation of the outcomes of basic research into advances in the field through Coordinated Agricultural Projects. A key to success in all of these areas has been the strict adherence to the guiding principles established at the start of the NPGI.

---

6   The National Academy of Sciences report entitled "Achievements of the National Plant Genome Initiative and New Horizons in Plant Biology", January 29, 2008. (http://www.nap.edu/catalog.php?record_id=12054).
7   Accessible at http://www.plantgenomesecrets.org/.

III. Plan for 2009-2013

NPGI Goal Statement

The goal of the NPGI is to develop a basic knowledge of the structures and functions of plant genes and translate this knowledge to a comprehensive understanding of all aspects of economically important plants and plant processes. By bridging basic research to plant performance in the field, the NPGI will accelerate basic discovery and innovation in economically important plants and enable enhanced management of agriculture, natural resources, and the environment to meet societal needs.

The new knowledge and insights gained from plant genomics are already leading to unexpected discoveries and conceptual advances in our understanding of the biology of plants. Through the advances made in the past 10 years, the scope of the NPGI has expanded beyond genomes to include large scale collections of proteins ("proteomes"), interactions between proteins ("interactomes"), metabolites ("metabolomes"), and collections of observable characteristics ("phenomes"). With these new tools, we can now develop a "systems" understanding of plants from the single cell through to the whole plant during development and under changing environmental conditions.

There has been an explosion of data from these diverse genome-scale approaches that will require a new level of analysis and integration if we are to understand fully the structure and function of plants and apply this knowledge to crop improvement. In the next five years, integration of multiple inputs to develop a "wiring diagram" for plants will be the first step towards the development of new kinds of crops. As the world increasingly relies on plants to provide solutions to the societal challenges of a growing need for plant-based fuels and materials as well as the impacts of climate change, continued innovation in all of these areas will be critical.

New Objectives for 2009-2013:

During the past decade of the NPGI, research has been moving from a focus on understanding plant genome structure and function towards a more integrated view of plants from the genome (the genetic blueprint of an organism) through to the phenome (the collection of its observable characteristics). The ability to do this has resulted from development of key resources, including whole genome sequences, RNA expression profiles, metabolite profiles, and proteomic data as well as the technology and tools needed to develop these resources and integrate them. There has been an explosive increase in the amount of data being generated through the NPGI and it has become a growing challenge to manage, store, analyze and integrate these diverse data types.

New solutions are needed that deal with today's issues, and that can accommodate and adapt to future needs. In addition, training of the current end-users from researcher to breeder, as well the next generation of scientists, is needed to insure that all can access and benefit from the wealth of available plant -omics resources. The objectives for the next five years of the NPGI will enable a diverse community of stakeholders to develop a more detailed, genome-enabled understanding of plants and translate this knowledge into new tools and resources.

Objective I. Expand genomic resources for every major plant of economic importance

In the past five years, sequencing and mapping efforts have focused on key genomes of economically important crops and their models. These efforts resulted in draft or completed reference genome sequences for Arabidopsis, rice, Medicago, poplar, maize, sorghum, soybean, and peach. In addition, other international efforts have contributed genome sequences for Brassica and grape, and additional genome sequences of varying extent are under development. The rapid advances in next-generation sequencing are opening up new possibilities for developing a complementary set of resources such as markers to meet end-user needs.

Understanding of plant epigenomes

In the first ten years of the NPGI, one focus area has been the decoding of sequences of key plant genomes. It is becoming evident that there is additional information carried by genomes that is not encoded in the sequence of bases but instead in the modifications that occur to the DNA and associated proteins that make up the chromosomes. This form of the genome, which can persist over multiple generations, is referred to as the "epigenome". The epigenome can control the differential expression of genes in specific cells and may change over the lifetime of an organism and in response to varying environmental conditions.

The economically important process of vernalization, central for the production of wheat and barley, is already known to be regulated epigenetically. Vernalization occurs when plants are exposed to prolonged low temperatures over winter, allowing them to flower the following spring. Development of new plant varieties that are adapted to changing climactic conditions will depend on an understanding of how epigenetic regulation occurs.

Plants represent a premier experimental system for understanding epigenetic regulation and have yielded much of our basic understanding of the impact of epigenetic modification on chromosome structure and gene expression. Capturing epigenomes and relating changes to plant growth and development is at the frontier of plant biology and will have broad impacts on future strategies for crop improvement. There will also be the need for technological breakthroughs toward new methods to capture epigenomes and enable assignment of functional significance to epigenetic modifications.

> *Epigenetic regulation can be a complex process and is already know to involve a series of "marks" on a genome that can include reversible modifications to the DNA itself as well as associated proteins such as histones. Epigenetic regulation has been implicated in a wide range of biological processes, including human diseases.*

Integrated comparative sequence resources

Although the cost of sequencing has come down dramatically in the past decade, it has been too expensive to sequence, finish, and annotate every plant genome to the quality of the Arabidopsis and rice genomes. However, reference genomes for the major plant groups of economic importance are enabling assembly of high quality draft sequences from close relatives. Comparative analysis of multiple genomes has been possible as more whole genome sequences have been completed. These integrated sequence resources are enabling a more detailed annotation of the reference genomes as well as providing clues about genome dynamics across members of major plant groups such as the grasses and the legumes. Comparative genome resources will continue to play an important role in understanding how gene networks evolved and function.

A new kind of reference genome

There are still several key plants of economic importance for which there is no reference genome and no whole genome sequence resources, including wheat, cotton, and pine. Several of the reference genomes that remain to be sequenced are large and complex. As sequencing costs decrease and next-generation sequencing methods mature, the choice of sequencing strategy and degree of coverage (for example, gene space versus whole genome) for the larger and more complex reference sequences will likely depend on the structure and organization of each genome and the utility to address specific scientific questions. These new reference resources may differ in degree and depth of sequence coverage from the Arabidopsis and rice genomes.

Survey sequence resources for thousands of plants

The increasing access to next-generation sequencing machines that can produce large amounts of sequence at relatively low cost has also opened up the opportunity to rapidly develop at low cost, additional sequence resources ranging from targeted regions and gene space to whole genome coverage. These resources will enable comparative studies to understand the evolutionary diversification of gene family structure and function, and provide more detailed information about genome dynamics across a plant clade. As sequencing costs continue to decrease, it should be possible to tackle thousands of plant genomes within the next five years. Continued coordination across the agencies participating in the NPGI will ensure seamless connection between technology development, development of new sequence resources, and the needs of the end-users.

Mining plant diversity

A major outcome of prior NPGI investments in structural genomics has been the recognition of the substantial genetic diversity within plant populations. This genetic diversity, which is captured in the germplasm resources for the key crops of economic importance in the U.S., serves as a repository of useful traits for overcoming biotic challenges such as disease and abiotic damage caused by lack of water and nutrients or temperature stress. The value of germplasm resources for plant research would be substantially increased by genotyping of the available resources for economically important plants such as soybean and maize. DNA markers called single nucleotide polymorphisms or "SNPs" will enable researchers to make plant breeding more efficient and precise, as well as accelerate breeding of crops enhanced for many key traits of interest.

The decreasing costs of sequencing now brings sequencing of collections of individual members of plant populations within reach. Within the next five years, it will be cost-effective to sequence individual plant genomes from a population to assess genetic diversity on a whole genome scale and apply this to breeding efforts.

Objective 2. Advance plant systems biology

The availability of a wealth of genome sequence resources and downstream resources for a core group of economically important plants and their models has opened up the possibility of understanding plants and their diverse interactions at a systems level. A full understanding of the processes controlling plant growth and development that are central to manipulating plant composition and yield requires more than DNA sequence information. Additional information about the gene regulatory sequences, RNAs, the proteins, metabolites, and other cellular components will be needed for the key crop plants and their models to develop a complete picture of the interplay between the structural and regulatory components that build plant cells and, in turn, whole plants in a time of rapid global change.

Toolkits to enable systems-level analysis of key plant processes

To date, much of the information regarding functional processes has been collected from samples containing multiple cells. New tools and methods are needed to dissect processes at a single cell level so that the molecules (mRNAs, proteins, metabolites, and so on) expressed in one or a few cells can be detected and quantified. This is particularly important to understand critical events that occur early on in processes such as embryogenesis and double fertilization, and in meristem development.

Understanding water use efficiency (WUE), nitrogen use efficiency (NUE) and nutrient utilization will enable development of new plant varieties that can withstand drought and require lower inputs of fertilizer. Progress in this area will require elucidation of the regulation of gene networks, proteins, and metabolites as well as tools to enable high-throughput capture of phenotypic information.

More efficient use of plant-based materials for biofuel production will be facilitated by development of feedstocks suited for this use. To develop these it will be important to understand the networks of genes involved in biomass characteristics and yield.

As trees may provide a long-term mechanism for sequestration of meaningful amounts of atmospheric carbon in terrestrial ecosystems, a full understanding of the gene networks involved in carbon cycling and plant interactions with the biotic and abiotic environment is needed to maximize this potential.

> *Fertilization in flowering plants is referred to as "double fertilization" because it involves two sperm nuclei. One nucleus fuses with the egg nucleus to form the diploid (2N) zygote from which the embryo develops. The second sperm nucleus fuses with two polar nuclei in the embryo sac to form the triploid (3N) primary endosperm nucleus. The endosperm provides nutrition for the developing embryo and is also an important source of nutrition for consumers of plant seeds.*

Regulation of plant structure and composition

Plants provide feedstock for many industrial products, including food, wood, paper, and increasingly, plant-based fuels. Development of new tools and resources to enable understanding of the pathways and processes that lead to changes in plant architecture will be needed to develop new plants with modified chemical and structural properties. Key traits will include those that impact vegetative development, wood formation, seed development, and specialized structures such as root nodules and trichomes, which can be used as biological "factories".

Objective 3. Translate basic discovery to the field

Development of new plant varieties with enhanced economic value depends on effective implementation of the advances in the basic understanding of plant growth and development. Critical needs include tools to enable rapid evaluation of traits under field conditions and to develop new varieties with valuable targeted traits. New plant varieties are needed to address challenges to agriculture brought about by climate change and to meet increased needs for plant-based materials and biofuels.

---

8   http://www.ontariocorn.org/classroom/products html

High-throughput phenotyping under field conditions

While plant phenotypes are being evaluated on a large scale under laboratory conditions, comparable evaluation under field conditions will require new approaches for capturing and analyzing data. Tools are needed for automated *in vivo* imaging, whole plant imaging, tracking, and analysis as well as integration with other environmental information about the plants under analysis.

Breeding for improved local adaptation to biotic and abiotic stress

Changing environmental conditions present new challenges for development of new crop varieties that can withstand a range of biotic and abiotic stress conditions including drought, heat, salinity, flooding, as well as pathogens that may have become more pathogenic through alterations of the disease triangle (the host, pathogen, and environment) or have been inadvertently introduced.

A National Genetic Trait Index

Development of a National Genetic Trait Index (NGTI) will facilitate the identification of useful genetic variation within the entire U.S. Germplasm Resources Information Network (GRIN) collection of plant accessions. The NGTI would contain information about key traits, including many of agronomic utility, for each plant. High-throughput genotyping and phenotyping will provide new knowledge and tools for breeders to use the nation's germplasm collections efficiently and to design new varieties that will meet producer and market needs.

Objective 4. Develop coordinated solutions to data access, data analysis and data synthesis

Plants are primary producers that are directly affected by climate change, yet have the capacity to mitigate consequences of such change. To utilize fully the growing knowledge of plant growth and development at scales ranging from the inside of a single cell to interactions of plants with the environment around them, it is necessary to understand how plants modulate responses to environmental variables, including light, temperature, water, and nutrients. A plant's environment is also impacted by other living organisms that provide nutrients, compete for resources, or serve as a source of stress through damage or disease. Use of genomic resources from specific organisms and metagenomic data for plant-associated organisms will greatly increase the understanding of the interplay between plants and their neighbors. Integration of data across all of these scales will form the basis for understanding the impact of genetic variation on plant growth and development and breeding of new crops suited to specific environments. Development of data standards and interoperability, as well as accessibility to primary data and metadata, will continue to be critical for data integration.

Data mining and visualization tools

As the amount and types of genomic data accumulate at an ever increasing rate, there is an urgent need for tools to serve a broad range of end-users from researchers to breeders. Challenges include functional genome annotation, comparative alignment across multiple whole genome sequences, integration of genome sequence information with other data types, and breeder-centric views of map and trait data that best serve their needs.

Modeling tools for predictive analysis

Development of a systems view of plants that integrates multiple types of data across several scales of analysis will be the first step towards a long-term goal of the development of predictive models. Using these models the impacts of changes in plant genotype and their interactions with changing environment can be assessed. For such a tool to be robust, it must be tightly connected to empirical observations so that it is part of an iterative cycle where experiments and models inform each other in an upward spiral.

> *A powerful but easy-to-use, internet-based, information management system is needed for the plant genetic resource community and the world's plant gene banks. The existing U.S. Germplasm Resources Information Network (GRIN) is being integrated into GRIN-Global to facilitate better conservation and understanding of plant genetic diversity.*

Objective 5. Enhance education, training and outreach

Research supported through the NPGI provides a unique opportunity to train the next generation of scientists across disciplinary boundaries that include the translation of basic discovery into improvement of existing crops and development of new economically important plants. The participation in the NPGI of agencies with missions that cover all aspects of plant research provides the potential to develop new training programs that build on the unique resources and expertise of each agency. It is no longer possible to take full advantage of all of the research advances and new tools and resources for plant biology without cross-disciplinary training and education.

Traineeships for undergraduates, graduate students and postdoctoral researchers to enable cross-disciplinary training

Training the next generation of plant researchers will be advanced by interdisciplinary training in key areas such as bioinformatics, quantitative genetics, and breeding. Plant genomics collaborations between university, government and industry laboratories will provide opportunities to develop a new cohort of researchers able to undertake and translate basic discoveries into applications.

Training opportunities for investigators at all levels

In order to enable every researcher in the Nation to take full advantage of advances in plant genomics research, the NPGI needs to continue to broaden participation of institutions, scientists, and students currently underrepresented in NPGI-supported activities. Opportunities are needed for both young and established researchers to add new skills to their toolboxes, such as bioinformatics, quantitative genetics, and plant breeding. These opportunities will strengthen their research, and in turn, the training they provide to their students in the laboratory, field, and classroom. In addition, inclusion of researchers from a broad range of institutions in training opportunities will broaden participation in NPGI-supported research.

Workshops to inform the broader research community about accessing and using NPGI research resources

The rich collection of tools and resources developed through the NPGI have been made more accessible to the community through workshops at major conferences organized by community databases such as GenBank, TAIR (The Arabidopsis Information Resource), MaizeGDB, and Gramene. In addition to providing training on how to deposit data and use the available resources, these workshops also provide an opportunity to promote participation in community-wide efforts to improve annotation of plant genomes and use new database contribution tools.

Outreach to the K-12 community

Recruitment of motivated students is critical to maintain a vibrant and productive plant genomics research community. Outreach to teachers provides an effective means of reaching large numbers of potential students. Professional development opportunities for teachers and trainees should be integrated into research programs to allow first-hand experience in cutting edge research and facilitate development of new teaching materials. There is significant need for outreach to schools that serve rural populations, members of underrepresented minorities, and the economically disadvantaged.

Communicating the outcomes of plant genome research to the public

The increasing dependence on plants for food, feed, and fuel makes it more important than ever to provide science-based information to the general public to enable informed decision making about such issues as genetically modified crops and the role of biotechnology in society. It is the role of every scientist involved in the NPGI to contribute accurate, science-based information about her or his work and its relevance to societal needs. In most cases, this communication should occur outside the realm of technical journals.

Objective 6. Broaden societal impacts

Effective interactions with industrial and international partners depends on linking basic discovery, development of new crops, and broadening the benefits deriving from the NPGI. It will be essential to foster strong connections through existing steering groups and develop additional interactions as new opportunities emerge.

Public-private partnerships

Public-private partnerships, involving industry, private foundations, and grower associations are integral to the continued success of the NPGI and avoiding redundant efforts. Private sector participation in the international rice genome sequencing project, maize genome mapping, and development of genotype resources for maize and soybean are examples of interactions that led to higher quality resources faster and more efficiently than would have been otherwise possible. The NPGI promotes participation of each entity as equal partners, with each contributing resources, openly sharing results with the broader community, and receiving appropriate credit.

International research coordination and cooperation

The Arabidopsis and rice genome sequencing projects were conducted by cooperating groups of international researchers that coordinated activities to ensure maximum efficiency and minimal duplication of effort. Other ongoing projects, including the Arabidopsis Functional Genomics Project and the Medicago and tomato sequencing projects, are using a similar approach. In each case the NPGI has followed the principle that each international partner should be supported by its own national resources and that coordination should be carried out by participating researchers on a scientific basis. The NPGI will continue to encourage NPGI-supported projects to take full advantage of opportunities for international coordination noting that future activities may include other types of –omics activities, as well as databases. In addition, the NPGI will continue to work with developing countries wherever possible to leverage resources and to promote greater cooperation toward solving problems of increasing need for food, fiber, and fuels.

IV. Appendix

The Interagency Working Group (IWG) on Plant Genomes solicited and collected input from many sources during the preparation of this report. The following are some of the workshops and reports that IWG considered in the development of the NPGI plan for 2009-2013.

- NSF workshop entitled "The National Plant Genome Initiative at Ten Years: A Community Workshop", Irvine, CA, August 26-28, 2008
- A White Paper entitled "Towards a Whole Genome Sequence of Common Bean, (Phaseolus vulgaris): Background, Approaches, Applications", August 2008.
- The National Academy of Sciences report entitled "Achievements of the National Plant Genome Initiative and New Horizons in Plant Biology", January 29, 2008. (http://www.nap.edu/catalog.php?record_id=12054).
- DOE Workshop on Carbon Cycling and Biosequestration, Office of Biological and Environmental Research, Office of Science, U.S. Department of Energy, March 2008
- NSF Plant Genome Research Program Awardee meeting discussions, September 2007 and September 2008.
- USDA-DOE Plant Feedstock Genomics for Bioenergy awardee meeting discussions, February 2007 and January 2008
- IWG Workshop entitled, "The National Plant Genome Initiative: What's Next?" held at the Plant and Animal Genome Conference, San Diego, CA, January 13, 2008.
- ERANET on Plant Genomics report on the forward look activity entitled "Plant Genomics Meets New Challenges", December 2007.
- National Wheat Genomics Conference Discussion on the Future of Wheat Genomics in the U.S., December 2007
- CSREES ASPB stakeholder workshop on Plant and Pest Biology, November 2007
- Strategic Opportunities for Cooperative Extension, National Association of State Universities and Land Grant Partners, October 2007
- CSREES Arachis Through Genomics and Biotechnology Stakeholder Workshop, October 2007
- CSREES Applied Grape Genomics Stakeholder Conference: Improved Plant Material and Diagnostic Techniques for the Future of Viticulture and Enology, University of California, Davis, July 16-17, 2007
- A White Paper: The International Barley Genome Sequencing Consortium (IBSC): A Coordinated Strategy for Sequence Analysis of the Barley Genome (Hordeum vulgare), (http://barleygenome.org)

- Formulation of the Organizational and Implementation Framework for the Global Partnership Initiative for Plant Breeding Capacity Building (GIPB) http://km.fao.org/gipb
- Rosaceae Specialty Crops Planning Workshop, June 2007
- Soybean Genomics Research, A Strategic Plan for 2008-2012, May 2007
- The Future of Maize Genetics: Planning for the Sequenced Genome Era, A Maize Genetics Community Retreat, March 2007.
- A White paper: Priorities for Research, Education and Extension in Genomics, Genetics, and Breeding of the Compositae, February 2007
- A Science Roadmap for Agriculture, National Association of State Universities and Land-Grant Colleges (NASULGC), May 2005 and update 2006
- Legumes as a Model Plant Family. Genomics for Food and Feed Report of the Cross-Legume Advances through Genomics Conference, December 2004
- The International Wheat Genome Sequencing Consortium, http://www.wheatgenome.org/
- Technology Roadmap Temperate Fruit Genomics, Genetics, and Breeding Workshop, October 2004
- A Workshop Report on Wheat Genome Sequencing, 2004

Acknowledgements

The Interagency Working Group on Plant genomes acknowledges the valuable contributions of former IWG members who participated during the last five years. Kay Simmons (U.S. Department of Agriculture) and Randy Johnson (U.S. Department of Agriculture) contributed to the preparation of the report. Lauren Kitchen (National Science Foundation) provided assistance in the production of the report.

Abstract

The National Plant Genome Initiative (NPGI) was established in 1998 as a coordinated national plant genome research program by the Interagency Working Group (IWG) on Plant Genomes with representatives from the U.S. Department of Agriculture (USDA), U.S. Department of Energy (DOE), National Institutes of Health (NIH), National Science Foundation (NSF), Office of Science and Technology Policy (OSTP), and the Office of Management and Budget (OMB). The current membership of the IWG also includes the U.S. Agency for International Development (USAID). Since 1998, the field of plant genomics has made tremendous strides under the NPGI. The NPGI has changed the way research is conducted in plant biology; it has attracted a new generation of scientists to plant research; and it has contributed new knowledge and ideas to science. Through the development of plant genomic resources, the NPGI has built a foundation on which the scientific community can advance research, not just in plant genomics but across diverse disciplines spanning the biological sciences to biotechnology.

In this report, the IWG describes the NPGI plan for the next five years (2009 – 2013). The IWG based this plan on a wide range of inputs from the broader community, which are listed in the Appendix. The objectives for the next five years will build on recent scientific and technical advances to ensure continued advancement in plant genomics specifically and plant sciences in general. In order to fully capitalize on the investments made to date and to enable further advances in plant genome research, the IWG recommends continued and increased investment in the NPGI by all participating agencies. The funds will continue to be expended on a competitive basis using rigorous peer review. As in the first ten years of the NPGI, the IWG will monitor progress and report major accomplishments annually.